ENGINEERING ETHICS: PEACE, JUSTICE, AND THE EARTH

Engineering Ethics: Peace, Justice, and the Earth

George D. Catalano

www.morganclaypool.com

1598290908   paper
1598290916   ebook

DOI   10.2200/S00039ED1V01Y200606ETS001

A Publication in the Morgan & Claypool Publishers' series
SYNTHESIS LECTURES ON ENGINEERING, TECHNOLOGY AND SOCIETY
Sequence in Series: Lecture #1

First Edition
10 9 8 7 6 5 4 3 2 1

Printed in the United States of America

ENGINEERING ETHICS: PEACE, JUSTICE, AND THE EARTH

George D. Catalano
State University of New York at Binghamton

SYNTHESIS LECTURES ON ENGINEERING, TECHNOLOGY AND SOCIETY #1

MORGAN & CLAYPOOL PUBLISHERS

ABSTRACT

A response of the engineering profession to the challenges of security, poverty and under-development, and environmental sustainability is described. Ethical codes, which govern the behavior of engineers, are examined from a historical perspective linking the prevailing codes to models of the natural world. A new ethical code based on a recently introduced model of Nature as an integral community is provided and discussed. Applications of the new code are described using a case study approach. With the ethical code based on an integral community in place, a new design algorithm is developed and also explored using case studies. Implications of the proposed changes in ethics and design on engineering education are considered.

KEYWORDS

Engineering ethics, models of the natural world, engineering design, engineering education

With gratitude and appreciation for my family and all my two and four legged friends and spiritual directors.

Contents

Preface

I have been a professor for nearly thirty years and have taught thousands of students who have pursued careers in engineering. Over the course of the last several decades, I have, as an engineer and an engineering professor, struggled with issues related to the environmental and societal impacts that technology has in the modern world. I have wondered what views of their responsibilities to society and the natural world do students take with them after graduation? Have I given them the tools to make their way in a world in which the natural world is under siege unlike any time before? How will they respond to the poverty and the injustices which dominate so much of our shrinking global society?

Over the course of my career, I have been a faculty member at colleges in the Deep South, the Midwest and now the Northeast, at large land grant institutions, elite military academies and small, predominantly liberal arts universities. I have also been a soldier during times of war. Much has changed in engineering education since my formal schooling where we imagined the engineering profession as value free. Today students do not let us get away with such a narrow view of engineering as more and more of them bring to the classroom an awareness of the state of the world's ecosystem as well as poverty and under-development throughout the world. We in engineering can no longer pretend that such issues are for some other profession, not ours.

My sincere hope in writing the present work is to provide a mechanism whereby issues related to the three great challenges that confront us today – security, poverty and under-development, and environmental sustainability – can be discussed in the context of the engineering profession.

I am grateful to many who have helped this effort become a reality. My sincerest thanks are extended to my family, my friends both two-legged and four-legged, and my many students. Each has played a part in the development of the ideas that I have put forward. I will remain forever in their debt. Thank you.

Pax et bene.

CHAPTER 1

Introduction

According to the Worldwatch Institute, in their annual report on the state of the world, we face three interrelated challenges: the challenge of security, including risks from weapons of mass destruction and terrorism, the challenge of poverty and under-development; and the challenge of environmental sustainability [1]. Technology and rapidly accelerating technical advances have played key roles in the creation of these challenges. Thus, engineers and the profession of engineering have much to say as to whether or not the challenges of security, poverty and sustainability can be successfully met.

To speak of a profession, particularly the profession of engineering, implies the following five characteristics, which are useful in distinguishing professions from nonprofessional occupations [2]. First, entrance into a profession requires a mastery of some set body of knowledge and thus involves an extensive period of intellectual training. Second, the professionals' knowledge and skills are seen as vital to the well being of the larger society. Third, professions typically have a monopoly or near monopoly on the provisions of their particular set of professional services. Fourth, professionals routinely have an unusual degree of autonomy in the workplace. Fifth, professionals claim to be regulated by ethical standards, usually embodied in a code of ethics. It is this last characteristic, the existence of ethical standards set forth in a code of conduct which is the focus of the present work.

1.1   THE CHALLENGE OF SECURITY

The challenge of security is at the forefront of everyone's attention today as it has been every day in the United States since the horrific event of September 11, 2001. That terrible tragedy as well as the 2004 terrorist attacks in Beslan in Russia [3], the bombing of trains in Madrid [4] on March 11, 2004 and many other terrorist attacks in Japan, Indonesia, the Middle east, other parts of Europe and elsewhere have all driven home the fact that we are not adequately prepared to deal with new threats. But better preparation may suggest a different kind of thinking or approach, not the traditional thinking from the past. This may be especially true for the profession of engineering.

As 2004 began, by one count, 24 significant on-going armed conflicts (1000 or more deaths) raged around the world, with another 38 hot spots that could slide into or revert to war. Armed conflicts have many costs, in addition to the cost in human lives that is reported in the news. The 1999 Report of the UN Secretary-General [5] put the economic costs to the international community of seven major wars in the 1990s, not including Kosovo, at $199 billion. In addition, there are the costs of economic losses to the countries actually at war. Other important areas affected by armed conflict are: eco-terrorism, environmental destruction as collateral damage, and social casualties.

Consider the impact of wars on children alone. A report prepared for the United Nations General Assembly reveals the full extent of children's involvement in the 30 or so armed conflicts raging around the world [5]. "Millions of children are caught up in conflicts in which they are not merely bystanders, but targets. Some fall victim to a general onslaught against civilians; others die as part of a calculated genocide. Still other children suffer the effects of sexual violence or the multiple deprivations of armed conflict that expose them to hunger or disease. Just as shocking, thousands of young people are cynically exploited as combatants."

According to Vesilind[1], engineering from its inception has been intimately associated with waging war. The earliest engineers were military engineers who worked at the behest of leaders who either were leading conquering armies or defending their conquered lands from invasion. Vesilund makes reference to the British linguist Young who in 1914 described the lineage of the word engineer tracing it back to the Latin word *ingenium*, an invention or engine [6]. Young adds, "There must have been confusion of Latin *ingenuus* and Latin *ingeniosus*. These should be opposite in meaning. I suppose an engineer ought to be ingenious and ingenuous, artful and artless, sophisticated and unsophisticated, bond and free." Vesilund concludes with a description of the dichotomy that he claims captures the essence of engineering today.

"The engineer is sophisticated in creating technology, but unsophisticated in understanding how this technology is to be used. As a result, engineers have historically been employed as hired guns, doing the bidding of both political rulers and wealthy corporations." [7]

1.2 THE CHALLENGE OF POVERTY AND UNDER-DEVELOPMENT

According to the Social Summit Programme of Action, "Poverty has various manifestations, including lack of income and productive resources sufficient to ensure sustainable livelihoods; hunger and malnutrition; ill health; limited access or lack of access to education and other basic services; increased morbidity and mortality from illness; homelessness and inadequate housing; unsafe environments; and social discrimination and exclusion. It is also characterized by a lack of participation

---

[1]P. Aarne Vesilind is Professor of Civil and Environmental Engineering and R.L. Rooke Chair in the Historical and Societal Context of Engineering in the Department of Civil and Environmental Engineering at Bucknell University. At Bucknell, Dr. Vesilind's teaching specialties are environmental engineering, and professional ethics among other areas.

in decision-making and in civil, social and cultural life. It occurs in all countries: as mass poverty in many developing countries, pockets of poverty amid wealth in developed countries, loss of livelihoods as a result of economic recession, sudden poverty as a result of disaster or conflict, the poverty of low-wage workers, and the utter destitution of people who fall outside family support systems, social institutions and safety nets." It further emphasizes that "Absolute poverty is a condition characterized by severe deprivation of basic human needs, including food, safe drinking water, sanitation facilities, health, shelter, education and information. It depends not only on income but also on access to social services." [8]

The challenge of poverty and under-development was addressed in October 2004 at the tenth anniversary of the International Conference on Population and Development. The United Nations International Conference on Population and Development (ICPD) was held from 5–13 September 1994 in Cairo, Egypt. During this two-week period, world leaders, high ranking officials, representatives of non-governmental organizations and United Nations agencies gathered to agree on a Program of Action [9]. The assembly found that significant progress in many fields important for human welfare has been made through national and international efforts. However, the developing countries are still facing serious economic difficulties and an unfavorable international economic environment, and people living in absolute poverty have increased in many countries.

Around the world many of the basic resources on which future generations will depend for their survival and well-being are being depleted and environmental degradation is intensifying, driven by unsustainable patterns of production and consumption, unprecedented growth in population, widespread and persistent poverty, and social and economic inequality. The main conclusions of the final report centered on the fact that while some progress has been made, poverty continues to undermine progress in many areas of the globe. Diseases such as HIV/AIDS are on the rise, creating public health time bombs in many counties. In addition, approximately 20 million children have died of preventable waterborne diseases. Today, hundreds of millions of people continue to live in squalor and

Population		Income
20%		82.7%
20%		11.7%
20%		2.3%
20%		1.9%
20%		1.4%

FIGURE 1.1: Distribution of world income, 1989 [10]

destitute conditions associated with the lack of clean drinking water and adequate sanitation.

The report recommends a series of principles [10] that focus on the challenges associated with poverty and underdevelopment in the world. Specifically, Principle 15 states:

"Sustained economic growth, in the context of sustainable development, and social progress require that growth be broadly based, offering
equal opportunities to all people."

The following figure (Fig. 1.1) from 1989 UNDP report, shows the unequal distribution of world income and illustrates the underlying reason why 3 billion people have virtually no recourse to the basic necessities of life.

The following table (Table 1.1) from UNDP (United Nations Development Program) shows the millions of people living without the basic necessities of food, education, water, and sanitation.

A fascinating comparison of actual expenditures on luxury items versus basic needs points to the issue of poverty and under-development as a result of priorities

TABLE 1.1: Priorities of Expenditures

REGION	ELIMINATING POVERTY: MASSIVE DEPRIVATION REMAINS, 2000 (MILLIONS)						
	LIVING ON LESS THAN $1 (PPP US$) A DAY	TOTAL POPULATION UNDER-NOURISHED[a]	PRIMARY AGE CHILDREN NOT IN SCHOOL	PRIMARY AGE GIRLS NOT IN SCHOOL	CHILDREN UNDER AGE FIVE DYING EACH YEAR	PEOPLE WITHOUT ACCESS IMPROVED WATER SOURCES	PEOPLE WITHOUT ACCESS TO ADEQUATE SANITATION
Sub-Saharan Africa	323	185	44	23	5	273	299
Arab States	8	34	7	4	1	42	51
East Asia and the Pacific	261	212	14	7	1	453	1,004
South Asia	432	312	32	21	4	225	944
Latin America and the Caribbean	56	53	2	1	0	72	121
Central & Eastern Europe & CIS	21	33	3	1	0	29	..
World	1,100	831	104	59	11	1,197	2,742

[a]1998–2000.

Source: World Bank 2003a, 2004f; UNESCO 2003; UN 2003.

is shown in Table 1.2. Compare the annual expenditure on makeup versus the funding provided for health care for all women or the money spent on perfumes versus the funds allocated to universal literacy for example.

With few exceptions, engineering and the engineering profession have not addressed issues related to poverty and under-development. Notably, Vallero describes engineering as becoming increasingly complex and the responsibilities of engineers to society as changing as society evolves [11]. According to Vallero, the great challenge in engineering is to treat all people fairly and justly though the motivation for doing so arises out of the observation that undesirable land use results in the creation of unhealthy environments which typically the least advantaged parts of society are forced to bear. Thus the responsibility to the poor is linked to the destruction of the natural environment not out of a sense of responsibility to address their specific conditions.

1.3   THE CHALLENGE OF ENVIRONMENTAL SUSTAINABILITY

The challenge of environmental sustainability results from a wide-ranging list of critical issues. A few examples of the urgency associated with sustainability are described. Concentrations of carbon dioxide, the main global warming gas in the Earth's atmosphere, posted the largest two-year increase ever recorded. Studies note that if the global temperature rises 2–6 degrees as now predicted, up to 35% of the world's species would become extinct by 2050 [12]. The United Nations published a study noting that the number of oceans and bays with "dead zones" of water, so devoid of oxygen that little life survives, has doubled to 146 since 1990 [13]. Two thirds of Caribbean coral reefs are threatened by human activities, including over-fishing and pollution runoff from agriculture. Toxic metals from discarded cell phones threaten both groundwater and he health of recyclers in Pakistan, India, China and elsewhere [14].

Polar ice caps and mountain glaciers are experiencing rapidly increasing temperature and as a result are melting at an accelerating rate. The Arctic Climate

TABLE 1.2: Annual Expenditure on Luxury Items Compared with Funding Needed to Meet Selected Basic Needs [10]

PRODUCT	ANNUAL EXPENDITURE (IN BILLION DOLLARS)	SOCIAL OR ECONOMIC GOAL	ADDITIONAL ANNUAL INVESTMENT NEEDED TO ACHIEVE GOAL (IN BILLION DOLLARS)
Makeup	18	Reproductive health care for all women	12
Pet food in Europe and United States	17	Elimination of hunger and malnutrition	19
Perfumes	15	Universal literacy	5
Ocean cruises	14	Clean drinking water for all	10
Ice cream in Europe	11	Immunizing every child	1.3

Impact Assessment study released in 2004 estimates that the Arctic Ocean might be ice-free by the end of the 21st century [15]. The ice sheets covering Greenland and Antarctica are seriously weakening and in many instances disintegrating. This disintegration would cause raising ocean levels around the world and concurrently significant flooding of coastal areas.

Over the course of the last few years, a team of U.S. and Canadian researchers said the Bering Sea was warming so much it was experiencing "a change from arctic to sub-arctic conditions." [16] Gray whales are heading north and walruses are starving, adrift on ice floes in water too deep for feeding. Warmer-water fish such as pollock and salmon are coming in, the researchers reported.

Off the coast of Nova Scotia, ice on Northumberland Strait has been so thin and unstable during the past few winters that thousands of gray seals crawled on unaccustomed islands to give birth. Storms and high tides washed 1500 newborn seal pups out to sea last year.

Focusing on one particular animal, the main natural habitat of the polar bear is under increasing threat as a consequence of the dramatic thinning of the Arctic sea ice. The link between the thinning of the ice and rising temperatures has been discovered by scientists at UCL and the Met Office Hadley Centre for Climate Prediction and Research, whose findings were published *Nature* [17]. The thinness of the ice covering the Arctic Ocean, approximately three meters deep, makes it far more vulnerable to longer summers than the glaciers of the Antarctic. A 40% thinning of the ice has occurred since the 1960s. Polar bears rely on the ice to hunt for seals, and its earlier break-up is giving them less time to hunt. Continued decrease in the Arctic's ice cover would also act to increase the effects of global warming in the northern hemisphere by decreasing the amount of sunlight reflected by the ice. It is also believed that the Arctic ice plays a role in the operation of the Gulf Stream, and that this could be disrupted by the continued thinning.

According to a new study funded by the Rockefeller Foundation, 75 percent of the farmland in sub-Saharan Africa is severely degraded and is being depleted of basic soil nutrients at an ominous rate, deepening a food production crisis that already affects 240 million people. The report, *Agricultural Production and Soil Nutrient Mining in Africa* [18], found substantial soil decline in every major region of sub-Saharan African, with the highest rates of depletion in Guinea, Congo, Angola, Rwanda, Burundi, and Uganda, where nutrient losses are more than 60 kilograms per hectare annually. The problem is exacerbated by fertilizer costs that are two to six times the world average, limiting the amount that African farmers can afford to buy, and the adoption of non-traditional farming techniques that sap the soil's fertility.

In summary, the world is in the midst of a period of unprecedented and disruptive change. This is particularly evident when examining the health of the world's ecological systems. A host of human forces impinge upon coral reefs, tropical rain forests and other critical natural systems located around the world. Half the planet's wetlands are gone. Total carbon emissions and atmospheric concentrations of carbon dioxide are both accelerating and 2004 was the fourth warmest year ever recorded. Over the course of the last 120 years, the ten warmest years have all occurred since 1990.

1.4   OTHER CHALLENGES

There are other challenges surfacing as we head forward into the 21st century that perhaps are a combination of the three challenges described previously. One issue that has recently gained considerable attention deals with the relation between internet service providers and China. The National People's Congress of the People's Republic of China (PRC) has passed an Internet censorship law in mainland China. In accordance with this law, several regulations were made by the PRC government, and a censorship system is implemented variously by provincial branches of state-owned ISPs, business companies, and organizations. The project is known as Golden Shield.

The operations of U.S. Internet companies in China are attracting concern in Congress after years of complaints from free speech and human rights advocates about these firms aiding Beijing's ability to censor content. Trade liberalization has sent China's economy booming, making it an attractive—even essential market for U.S. companies to enter. But China's government has retained tight political controls. China is believed to have the world's most sophisticated network for monitoring and limiting information online. Combined with Beijing's controls on traditional media, this surveillance program has limited domestic debate on issues such as China's human rights record, Tibetan independence, or Taiwan. Hardware and software provided by many major U.S. technology companies have allowed the PRC government to use the internet as a tool for controlling public opinion.

1.5   CONCLUDING REMARKS

The question then becomes how can the engineering profession effectively respond to these three challenges, that is, to the challenge of peace, of poverty and under-development, and of environmental sustainability? As responses will depend upon the underlying ethical foundation for the engineering profession as described in the ethical codes, the next section of this work will examine engineering codes of conduct and seek to place them in an historical context.

C H A P T E R   2

Engineering Ethics

Engineering applies technical knowledge to solve human problems. More completely, engineering is a technological activity that uses professional imagination, judgment, integrity, and intellectual discipline in the application of science, technology, mathematics, and practical experience to design, produce, and operate useful objects or processes that meet the needs and desires of a client [19]. Today engineering is seen as a profession which refers specifically to fields that require extensive study and mastery of specialized knowledge and a voluntary and abiding commitment to a code of conduct which prescribes ethical behavior.

2.1   HISTORICAL OVERVIEW

The academic discipline of ethics, also called moral philosophy, involves arranging, defending, and recommending concepts of right and wrong behavior [20]. Philosophers today divide ethical theories into three general subject areas (Fig. 2.1): meta-ethics, normative ethics, and applied ethics. Meta-ethics explores the origins of our ethical standards. Normative ethics seeks to provide standards that can govern right and wrong behavior. Applied ethics focus on specific issues such as, for example, the existence of huge nuclear arsenals in the U.S. and the former Soviet Union, the existence of huge debts among countries in the Third World debt, and the treatment of animals on factory farms. In engineering, codes of conduct, developed to regulate the behavior of the practicing engineer, are examples of normative ethics, which shall be the focus of this review.

FIGURE 2.1: Subject areas in ethics

Normative ethics (Fig. 2.2) involves arriving at moral standards that regulate right and wrong conduct. The Golden Rule [21] is an example of such a moral standard. The key assumption in normative ethics is that there is only one ultimate criterion of moral conduct, whether it is a single rule or a set of principles. Three variations in normative ethics are: (1) virtue theories, (2) duty theories, and (3) consequentialist theories [22].

Virtue ethics places less emphasis on learning rules, and instead stresses the importance of developing good habits of character. Historically, virtue theory is one of the oldest normative traditions in Western philosophy, having its roots in ancient Greek civilization. Plato[1] emphasized four virtues in particular, which were later called *cardinal virtues*: wisdom, courage, temperance and justice.

Duty theories base morality on specific, foundational principles of obligation. Two terms that are used to describe duty theories are deontological from the Greek word *deon* or duty and nonconsequentialist as these principles are

---

[1]Plato, 429–347 B.C.E., is considered by most scholars to be one of the most important writers in the Western literary tradition and one of the most penetrating, wide-ranging, and influential authors in the history of philosophy. An Athenian citizen of high status, he displays in his works his absorption in the political events and intellectual movements of his time, but the questions he raises are so profound and the strategies he uses for tackling them so richly suggestive and provocative that educated readers of nearly every period have in some way been influenced by him. In practically every age since the Ancient Greek civilization, there have been philosophers who continue the ideas of Plato and adopt the description as Platonists in some important respects.

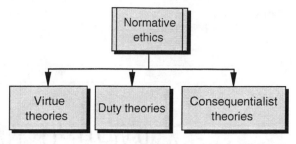

FIGURE 2.2: Theories of normative ethics

obligatory, irrespective of the consequences that might follow from our actions. Kant's categorical imperative [23] is an example of a duty-based theory of ethics. Kant[2] gives at least four versions of the categorical imperative, but one is especially direct: Treat people as an end, and never as a means to an end. That is, we should always treat people with dignity, and never use them as mere instruments. Ross[3] suggested in a more recent theory that there exists the following duties in living an ethical life [24]:

- Fidelity: the duty to keep promises.

- Reparation: the duty to compensate others when we harm them.

- Gratitude: the duty to thank those who help us.

- Justice: the duty to recognize merit.

- Beneficence: the duty to improve the conditions of others.

- Self-improvement: the duty to improve our virtue and intelligence.

- Non-maleficence: the duty to not injure others.

[2]Immanuel Kant, 1724–1804, is often considered one of the greatest, and most influential thinkers of modern Europe and the last major philosopher of the Enlightenment. His most original contribution to philosophy is his "Copernican Revolution," that it is the representation that makes the object possible rather than the object that makes the representation possible. This introduced the human mind as an active originator of experience rather than just a passive recipient of perception.
[3]W.D. Ross was a Provost of Oriel College, Oxford, Honorary Fellow of Merton College and a Fellow of the British Academy during the early 20th century. His greatest contribution was found in his criticisms of consequentialist moral theories.

FIGURE 2.3: Theories of consequentialism

Consequentialist theories (Fig. 2.3) or principles of ethical behavior require we do a kind of cost-benefits analysis to ultimately decide whether or not an action is ethical or unethical. Consequentialism requires that counting or estimating both the good and bad consequences of an action, and determining whether the total good consequences outweigh the total bad consequences. The action is morally proper if the good consequences outnumber the bad consequences. If the bad consequences are greater, then the action is morally improper. Three subdivisions of consequentialism [25] are:

- Ethical Egoism: an action is morally right if the consequences of that action are more favorable than unfavorable only to the agent performing the action.

- Ethical Altruism: an action is morally right if the consequences of that action are more favorable than unfavorable to everyone except the agent.

- Utilitarianism (Fig. 2.4): an action is morally right if the consequences of that action are more favorable than unfavorable to everyone.

FIGURE 2.4: Theories of utilitarianism

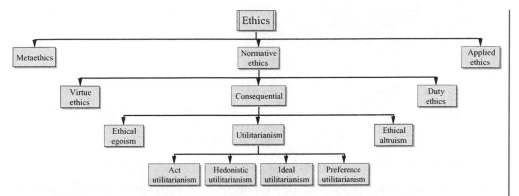

FIGURE 2.5: Overview of ethics with connections to engineering codes

It is Utilitarianism which has had a significant impact on the development of ethical codes of behavior within the engineering profession as can be observed in a careful reading of the different engineering codes (Fig. 2.5). There have been different variations of utilitarianism developed. Bentham[4] suggested two kinds: act-utilitarianism (i.e. tally the consequences of each action and thereby determine on a case by case basis whether an action is morally right or wrong) and hedonistic utilitarianism (i.e. only pleasurable consequences matter in the calculation) [26]. Subsequent versions are identified as rule utilitarianism, ideal utilitarianism and preference utilitarianism. Moore[5] proposed ideal utilitarianism, which involves tallying any consequence that is intuitively recognized as good or bad rather than as causing pleasure or pain [27]. Hare[6] proposed preference utilitarianism, which involves tallying any consequence that fulfills our preferences.

[4]Jeremy Bentham, February 15, 1748–June 6, 1832, was an English jurist, philosopher, and legal and social reformer. He is best known as an early advocate of utilitarianism and animal rights, who influenced the development of liberalism.

[5]G.E. Moore, 1873–1959, was a highly influential British philosopher of the early twentieth century. Moore's main contributions to philosophy were in the areas of metaphysics, epistemology, ethics, and philosophical methodology.

[6]Richard Mervyn Hare, March 21, 1919–January 29, 2002, was an English moral philosopher, who held the post of White's Professor of Moral Philosophy at the University of Oxford from 1966 until 1983, and then taught for a number of years at the University of Florida. Peter Singer, known for his work in animal liberation, was also a student of Hare's, and has explicitly adopted many elements of Hare's thought.

2.2 REVIEWING TODAY'S CODES OF ETHICS

At the start of the 21st century, there are as many different codes of conduct in engineering as there are engineering disciplines and specialties. One professional society, the National Society for Professional Engineers (NSPE), has offered one general code which is widely employed today in all the disciplines as well as in engineering education. The NSPE Code of Ethics consists of a preamble followed by a listing of fundamental canons and then rules of practice [28]. The very first canon cautions engineers in the fulfillment of their professional duties, to "hold paramount the safety, health and welfare of the public." As a result, the first rule of practice states that engineers shall "hold paramount the safety, health, and welfare of the public." Note that the explicit requirements focus on the public though there is no indication as to who is considered to be part of the public. Nor does the code refer to any of the challenges outlined as critical in the previous section. There is no indication that peace and security ought to be considered or issues related to poverty and the under-developed world nor environmental sustainability.

The American Society of Mechanical Engineers (ASME) sets forth a similarly constructed code of ethics with fundamental principles followed by fundamental canons [29]. The first principle states that engineers uphold and advance the integrity, honor, and dignity of the Engineering profession by using their knowledge and skill for the enhancement of human welfare. The supportive fundamental canon states engineers shall hold paramount the safety, health and welfare of the public in the performance of their professional duties.

The American Society of Civil Engineers (ASCE) does at least mention the environment in its code [30]. According to ASCE, Engineers uphold and advance the integrity, honor and dignity of the engineering profession by using their knowledge and skill for the enhancement of human welfare and the environment (fundamental principle) and shall hold paramount the safety, health and welfare of the public and shall strive to comply with the principles of sustainable development in the performance of their professional duties (fundamental canon). There is no explanation of what is meant by the enhancement of the environment. In November

1996, the ASCE Board of Direction adopted the following definition of sustainable development: "Sustainable development is the challenge of meeting human needs for natural resources, industrial products, energy, food, transportation, shelter, and effective waste management while conserving and protecting environmental quality and the natural resource base essential for future development." [31]

The Institute of Electrical and Electronics Engineers (IEEE) Code of Ethics states that its members accept responsibility in making engineering decisions consistent with the safety, health and welfare of the public, and to disclose promptly factors that might endanger the public or the environment [32]. Here, an interesting notion of responsibility towards the environment is described. It is not in opposition to the IEEE code to endanger the public or the environment only to not disclose promptly factors that might endanger the public or the environment.

The Institute of Industrial Engineers [33] (IIE) endorses the Canon of Ethics provided by the Accreditation Board for Engineering and Technology (ABET) whose first principle is that engineers uphold and advance the integrity, honor and dignity of the engineering profession by using their knowledge and skill for the enhancement of human welfare and whose first canon is engineers shall hold paramount the safety, health and welfare of the public in the performance of their professional duties [34]. ABET is the accrediting body for all engineering and engineering technology programs in the United States and thus has an important impact on the training of tomorrow's engineers and engineering educators.

Members of the American Institute of Chemical Engineers (AIChE) are challenged to uphold and advance the integrity, honor and dignity of the engineering profession by being honest and impartial and serving with fidelity their employers, their clients, and the public; striving to increase the competence and prestige of the engineering profession; and using their knowledge and skill for the enhancement of human welfare [35]. To achieve these goals, AIChE members shall hold paramount the safety, health and welfare of the public and protect the environment in performance of their professional duties. There is neither elaboration on the idea of protecting the environment nor an identification on from whom or what shall it be protected.

Many other engineering disciplines exist, each with their own codes for ethical conduct. As can be seen from this review, a large percentage of the codes do not explicitly identify the environment as an important stakeholder in discussions of the ethics of engineering choices. Equally as troubling, those codes that do mention the environment refer to the idea of enhancing nature or promoting sustainable development, which is based solely upon meeting human needs. A select few number of codes do mention a responsibility to protect the environment but without identifying from whom or from what. There is no reference to the challenge of peace and security. In addition, there is no reference implicit or explicit to the challenge of poverty and the under-developed world. A comparison of many different codes is presented in Table 2.1.

There are many other engineering disciplines at present, each with its own code of conduct or ethics, which describes the responsibilities of the profession. Most focus heavily on the sense of responsibility engineering has towards employers, society in general and towards other professional engineers.

2.3 CONCLUDING REMARKS

Engineering as a value-laden profession seeks to codify ethical behavior with various codes of conduct as put forth by different engineering societies. There are differences among the different codes but there are some striking similarities. The similarities exist in what has not been included in the ethical codes. While each does speak to the importance of holding paramount the public safety, issues associated with the intimate connection between engineering and war industries and terrorism are not discussed. In addition, no code speaks to the challenge of world poverty and the plight of the under-developed world. With one exception, that of ASCE, the challenge of environmental sustainability is completely ignored.

If we as engineers are to face these important challenges of our time, it may require a significantly different ethical code, one that can only come about if we

TABLE 2.1: Comparison of Attitudes Towards Security, Poverty and Nature among Various Engineering Societies

CODE OF CONDUCT	RELEVANT CANONS AND PRINCIPLES	ATTITUDES TOWARDS SECURITY	ATTITUDES TOWARDS POVERTY	ATTITUDES TOWARDS NATURE
NSPE	Hold paramount the safety, health, and welfare of the public.	No explicit reference	No explicit reference	No explicit reference
ASME	Uphold and advance the integrity, honor, and dignity by using their knowledge and skill for the enhancement of human welfare.	No explicit reference	No explicit reference	No explicit reference
ASCE	Hold paramount the safety, health and welfare of the public and shall strive to comply with the principles of sustainable development.	No explicit reference	No explicit reference	Sustainable development linked solely to meeting human needs
IIE	Accept responsibility in decisions consistent with the safety, health and welfare of the public, and to disclose promptly factors that might endanger the public or the environment.	No explicit reference	No explicit reference	Endangering environment not explored
IIE (ABET)	Shall hold paramount the safety, health and welfare of the public in the performance of their professional duties.	No explicit reference	No explicit reference	No explicit reference
AIChE	Hold paramount the safety, health and welfare of the public and protect the environment.	No explicit reference	No explicit reference	Protecting the environment not explored

view our professional responsibilities in a much broader way. The question then becomes how can we view our sense of ethics in a different way?

The era of modern engineering begin during the Renaissance and flourished as a result of the Industrial Revolution during the 18th and 19th centuries. One of the most important concurrent developments in ethical theory is termed Utilitarianism. Utilitarianism (from the Latin *utilis*, useful) is a theory of ethics that prescribes the quantitative maximization of good consequences for a population [36]. It is a single value system and a form of consequentialism and absolutism. This good to be maximized is usually happiness, pleasure, or preference satisfaction. Engineers are by and large utilitarians, seeking to maximum the good that is done [37]. Engineering codes of conduct demonstrate the strong influence of utilitarianism on our sense of responsibility. While utilitarianism has been useful in developing our sense of ethics within engineering, it has not permitted us to broaden that sense of responsibility to include peace and security, the challenge of poverty and the under-developed world and of environmental sustainability.

Utilitarianism was developed at a time when the world was imagined to be a machine which ultimately could be analyzed and divided into its many parts. Nature, the Earth, the Universe were all thought to be governed by immutable laws which we sought to uncover. Yet science has changed dramatically since the era of the mechanical universe. Perhaps we need to examine our sense of ethical responsibility in light of a newer scientific paradigm, one more indicative of the science of the 21st century rather than the 18th and 19th centuries.

C H A P T E R   3

Models of the Earth

Various models have been proposed to describe the behavior of the Earth and the Universe. Such models have changed as the scientific paradigm of the time has changed. We shall examine several of the models proposed beginning with the Middle Ages in Western Europe and ending with one of the most current models used today at the outset of the 21st century.

3.1   EARTH AS GREAT CHAIN OF BEING

Modern engineering in many respects begin with the Renaissance period in Western Europe. Humankind's understanding of or model for the natural world radically shifted from the notion of the Great Chain of Being [38] prevalent during the medieval period to the Universe as a mechanical clock [39]. The Great Chain of Being, reproduced in Fig. 3.1, is a powerful visual metaphor for a divinely inspired universal hierarchy ranking all forms of higher and lower life; the male alone represents humans. The top of the chain represents perfection in the highest degree. Most believers in the chain call this God. The chain in its entirety represents all degrees of perfection from the highest and fullest to the lowest and least; it is complete. Hence the universe would not be complete if the chain did not extend all the way to the bottom or if it had gaps in it. The universe is more perfect (in the sense that it is more complete) if all degrees of perfection are represented in it

FIGURE 3.1: Great chain of being a powerful visual metaphor for a divinely inspired universal hierarchy ranking all forms of higher and lower life; the male alone represents humans. From Didacus Valades, Rhetorica Christiana (1579). Reproduced here from Anthony Fletcher's Gender, Sex, & Subordination [38]

than if only the highest is represented. Eco[1] described this paradigm as "a place for everything and everything in its place." [40]

3.2   EARTH AS MECHANICAL CLOCK

Galileo is credited with being responsible for one of the most significant revolutions in thought in the development of the Western world and is referred to by many scholars as the originator of experimental science. In fact the term "revolution" was coined in response to his opinion that our Earth was one of several planets that revolved around the sun [41]. The work of Galileo[2] as well as many other natural philosophers led to the Age of Enlightenment, which refers to the 18th century in European philosophy, and is often thought of as part of a larger period, which includes the Age of Reason. Within the "enlightenment" movement, rationality was advocated as a means to establish an authoritative system of ethics, aesthetics, and knowledge. The intellectual leaders of this movement regarded themselves as courageous and elite, and regarded their purpose as leading the world toward progress and out of a long period of doubtful tradition, full of irrationality, superstition, and tyranny which they believe characterized the medieval period. Nature possessed only instrumental value and thus must be not only managed or controlled but transformed into useful resources that fed the insatiable appetites of progress. The visual metaphor used to depict the natural world became the mechanical clock (Fig. 3.2) rather than the Great Chain of Being.

There was a second revolution, a moral one, which resulted from Galileo's findings. Man was believed to have been made in God's image, was the completion and moral center of the created world. If Galileo was right, man held an insignificant position in the physical universe which did not seems fitting for the moral center

[1]Umberto Eco, born January 5, 1932, is an Italian medievalist, philosopher and novelist, best known for his novel *The Name of the Rose* and his many essays.
[2]Galileo Galilei, February 15, 1564–January 8, 1642, was an Italian physicist, astronomer, and philosopher who is closely associated with the scientific revolution. His achievements include improvements to the telescope, a variety of astronomical observations, the first and second laws of motion, and effective support for Copernicanism.

FIGURE 3.2: Mechanical clock as metaphor for nature (The Tower Clock of the Cathedral Church of St. James, Toronto)

of the universe. Physical centrality was understood to signify moral centrality, and Galileo appeared to be denigrating the dignity of man and thereby denying God's scheme of values.

The belief that humanity is the moral center of the universe has had lasting endurance. There have been dissenting voices but the predominant view has been that humans are, rightly, of overriding or exclusive moral significance. In our actions regarding the nonhuman world, we have usually only been concerned with human values. Our goals, our technologies, have focused on how best to utilize the natural world to benefit humans. Various religious and secular reasons have been given for our pre-eminent moral standing. Humans unlike the lower animals are said to have souls or to be morally superior because of their rationality. Or humans are said to deserve our privileged position because of our seemingly victorious evolutionary struggle or simply because we have made up the rules.

Consider the attitudes towards the natural world expressed by the various engineering societies as listed in Table 2.1. Clearly, those attitudes originated in the scientific and philosophical theories of the Age of Enlightenment. Engineers should care about nature if at all only if it serves the interests of humankind. Nature has

FIGURE 3.3: The Gaia hypothesis: Earth as a living system

no intrinsic value only instrumental value. Nature needs to be managed, controlled and manipulated to serve us.

3.3   EARTH AS LIVING SYSTEM

Other metaphors have been used to model the natural world. One which had some notoriety in the late 20th century was the Gaia hypothesis [42]. In 1965, Lovelock[3] published the first scientific paper suggesting the Gaia hypothesis. The Gaia hypothesis, Fig. 3.3, states that the temperature and composition of the Earth's surface are actively controlled by life on the planet [43]. It suggests that if changes in the gas composition, temperature or oxidation state of the Earth are caused by extra-terrestrial, biological, geological, or other disturbances, life responds to these changes by modifying the abiotic environment through growth and metabolism. In simpler terms, biological responses tend to regulate the state of the Earth's environment in their favor.

In support of the Gaia hypothesis, Margulis[4] first put forward her creative theory of endosymbiosis. Endosymbiosis attempts to specify the relationship between organisms which live one within another in a mutually beneficial relationship with one serving as a host cell (the boss cell) and another the symbiont (the

[3]James Lovelock, born July 26th, 1919, is an independent scientist, environmentalist, author and researcher, He is the author of "The Gaia Theory", and "The Ages of Gaia", which consider the planet Earth as a self-regulated living being.

[4]Lynn Margulis is a Distinguished University Professor at the University of Massachusetts. She has made original contributions to cell biology and microbial evolution to developing science teaching materials and hands-on garbage and trash projects in elementary schools.

dependent organism) which resides within the host cell. As was the case when the Gaia Hypothesis was first outlined by Lovelock, the concept of endosymbiosis was so new and required such a degree of leading edge specialised information, that it was often completely misunderstood—not only by researchers in unrelated fields, but also by her peers.

3.4   EARTH AS SELF-ORGANIZING SYSTEM

Modern science at the start of the 21st century does not model the natural world using either the great chain of being or the mechanical clock paradigms or as living being (Gaia hypothesis). Today the natural world is most often described using the model of a self-organizing system and nature rather than being thought of as immutable is seen as constantly in change (Fig. 3.4). Self-organization refers to a process in which the internal organization of a system, normally an open system, increases automatically without being guided or managed by an outside source [44]. Self-organizing systems typically (though not always) display emergent properties. Emergence is the process of complex pattern formation from simpler rules [45]. This can be a dynamic process (occurring over time), such as the evolution of the human brain over thousands of successive generations; or emergence can happen over disparate size scales, such as the interactions between a macroscopic number of neurons producing a human brain capable of thought (even though the constituent

FIGURE 3.4: An example of a self-organized system: The world's weather patterns (NOAA, December–January 1991) [47]

neurons are not themselves conscious). For a phenomenon to be termed emergent it should generally be unpredictable from a lower level description.

The world abounds with systems and organisms that maintain a high internal energy and organization in seeming defiance of the laws of physics [46]. According to Decker, "As a bar of iron cools, ferromagnetic particles magnetically align themselves with their neighbors until the entire bar is highly organized. Water particles suspended in air form clouds. An ant grows from a single-celled zygote into a complex multicellular organism, and then participates in a structured hive society. What is so fascinating is that the organization seems to emerge spontaneously from disordered conditions, and it does not appear to be driven solely by known physical laws. Somehow, the order arises from the multitude of interactions among the simple parts. The laws that may govern this self-organizing behavior are not well understood, if they exist at all. It is clear, though, that the process is nonlinear, using positive and negative feedback loops among components at the lowest level of the system, and between them and the structures that form at higher levels."

Decker goes on to add, "The study of landscape ecology provides an example of how a self-organizing system or SOS perspective differs from standard approaches. Ecologists are interested in how spatial and temporal patterns such as patches, boundaries, cycles, and succession arise in complex, heterogeneous communities. Early models of pattern formation use a 'top-down' approach, meaning the parameters describe the higher hierarchical levels of the system. For instance, individual trees are not described explicitly, but patches of trees are. Or predators are modeled as a homogenous population that uniformly impacts a homogeneous prey population. In this way, the population dynamics are defined at the higher level of the population, rather than being the results of activity at the lower level of the individual."

Decker argues that the problem with this top-down approach is that it violates two basic features of biological phenomena: individuality and locality. By modeling a pack of wolves as a amorphous, homogeneous mob there is no allowance for old wolves, young pups, sick wolves or aunts and uncles who spend most of their time pack-sitting. Nor is there allowance for expert hunters or, conversely, young and inexperienced hunters.

These differences can lead to larger differences—such as changes in the population gene frequencies, individual body size, or population densities—that might have cascading effects at still higher levels. The principle of locality means that every event or interaction has some location and some range of effect. Decker states, "To say that a system is self-organized is to say it is not entirely directed by top-down rules, although there might be global constraints on the system. Instead, the local actions and interactions of individuals generate ordered structures at higher levels with recognizable dynamics. Since the origins of order in SOS are the subtle differences among components and the interactions among them, system dynamics cannot usually be understood by decomposing the system into its constituent parts. Thus the study of SOS is synthetic rather than analytic."

If the self-organized system is used to model the natural world rather than the great chain of being or the mechanical clock, our sense of responsibilities to the natural world seem to change significantly. We are forced to look "synthetically" rather than "locally," that is at the very least or moral sphere of concern must broaden. Secondly, nature is no longer in perfect order nor is it a collection of parts (i.e. gears, levers, weight) which can be replaced or modified according to our desires. The mechanical clock in many ways has been replaced by a seemingly chaotic clock which defies predictability, single-valueness and repeatability. If we are to make sense of our place in this natural world, we need a very different sense of ethics. One attempt at providing such an ethical framework has been offered by Johnson in his development of a morally deep world [48].

3.5   CONCLUDING REMARKS

Models that have been proposed to help in our understanding of Nature and the Universe and our place within have changed many times over the course of the last two thousand years. Nature was imagined once as a great chain of being, with everything in its place and a place for everything. Then with the work of Newton and others, Nature became a machine, losing its enchantment, existing only in material form. Eventually that model was also supplanted by a notion of Nature or the Earth a living organism. Recently, with developments in the science of chaos and complexity, a new model for the Universe has been offered, that is, Nature as a self-organizing system. One thing is clear—there will be more models very soon on our horizon.

CHAPTER 4

Engineering in a Morally Deep World

A new approach to engineering ethics is developed, one based on the notion of a morally deep world. The morally deep world was first developed within the context of environmental ethics. A key element in its development in environmental ethics is the identification of an integral community. The present chapter makes the case for extending the identified integral community to include not only the environment but also other segments of society which have not been included in engineering ethics cases in the past.

4.1 BORROWING FROM ENVIRONMENTAL ETHICS

In A Sand County Almanac [49], Leopold[1] declares: "A thing is right when it tends to preserve the integrity, stability, and beauty of the biotic community. It is wrong when it tends otherwise." According to Leopold, acting ethically is a matter of concern both for us and for others with whom we are in some sort of community. The notion of a community deserves some discussion. We perhaps are most comfortable with community referring to a body of people having common

[1]Aldo Leopold, 1886–1948, was an American naturalist, conservationist, and philosopher of profound importance to the environmental movement. His most significant contribution was the development of the Land Ethic. Leopold wrote, "[A] land ethic changes the role of Homo Sapiens from conqueror of the land-community to plain member and citizen of it. It implies respect for his fellow-members, and also respect for the community as such."

rights, privileges, or interests, or living in the same place under the same laws and regulations; as, a community of Franciscan monks. In biology or ecology, community refers to an interacting group of various species in a common location. For example, a forest of trees and undergrowth plants, inhabited by animals and rooted in soil containing bacteria and fungi, constitutes an integral community. Extending the notion of community in this way is consistent with the pattern evidenced in human society over the centuries. We have progressively enlarged the boundaries of our understanding of community and recognized the membership of slaves, foreigners, etc., those for whom membership was not extended at earlier times in history. Leopold's land ethic then "simply enlarges the boundaries of the community to include soils, waters, plants, and animals, or collectively: the land."

Johnson[2] discusses how non-sentient land can count morally and focuses upon the concept of a living being. [50] For Johnson, a living being is best thought of not as a thing of some sort but as a living system, an ongoing life-process. A life-process has a character significantly different from those of other processes such as thermodynamics processes for example. Our character, as living beings, is the fundamental determinant of our interests. Johnson adds further that:

> The interests of a being lie in whatever contributes to its coherent effective functioning as an on-going life-process. That which tends to the contrary is against its interests. . . . moral consideration must be given to the interests of all living beings, in proportion to the interest. Some living systems other than individual organisms are living entities with morally considerable interests. . . . All interests must be taken into account.

A shift to a morally deep world-view in engineering would have a profound impact on the sense of ethical responsibility that the engineering profession would embrace. The next section will provide several examples of application of a morally

---

[2]Lawrence Johnson is a philosopher on the faculty at Flinders University, South Australia. He advocates a major change in humankind's attitudes towards the non-human world.

deep code. Prior to examining those cases, which directly focus upon engineering, I may be useful to consider first a case study involving the environment. The particular case to be examined involves a restoration of a predator, the wolf, into an ecosystem, the southwestern part of the United States.

4.2   CASE STUDY 1: WOLVES IN THE SOUTHWESTERN U.S.

Let us consider the implications of the morally deep argument for the case of restoration of wolves into the Southwest, one of the most contentious issues in wildlife management today [51]. For the purposes of illustration, let us focus on the land near the White Sands Missile Range near Las Cruces, New Mexico. Johnson would challenge us to first identify all the members of the community. For this example a listing would include the following:

- Wolves

- Prey animals including domestic sheep and cattle as well as deer, rabbits, coyotes, and others

- Desert lands

- Ranchers and sheep farmers

- Hunters

- US Fish and Wildlife Service and other state and local government agencies

- US Department of Defense

- Residents of White Sands and nearby towns and settlements

- Residents of New Mexico and the entire United States

- Native American residents

Often in such cases, two very different perspectives dominate the deliberations. On one side of the debate is atomism, a view that moral assessment applies only to individuals. The individual would be individual wolves, prey, ranchers, etc.

On the other side is holism, a view that collectives or whole are subject to moral appraisal. In a morally deep world, the view is shortsighted morally if one adopts either a holistic or atomistic. No one (holistic or atomistic) interest has priority over the other. There is an inevitable tension between atomistic and holistic ethics. Sometime the interests of the biotic community will outweigh the interests of the individual, while at other times it is the interests of the individual, which are paramount. Let us next identify the extent of the community or living being in this case. Recall that a living being is characterized as having an ongoing life process with interests in whatever contributes to its coherent effective functioning. Clearly wolves, their prey, the desert lands, ranchers, sheep farmers, hunters and people who live in or near White Sands have considerable interests. Other identified elements could be argued to have less interest in the coherent effective functioning of the community. That is not to suggest that, for example, the residents in New York would have no interest in the restoration but their impact on the coherent effective functioning of the ongoing process would be less.

An interesting example of the tension between atomism and holism can be identified in the following scenario. Suppose wolves are restored to the White Sands Missile range desert and suppose that, as has been the case in Yellowstone National Park, wolves adapt well and quickly grow in numbers [52]. In Yellowstone, some wolves are routinely killed as part of wolf or game management practices [53]. From a holistic perspective this may be morally acceptable but it would be difficult to justify the killing from an atomistic perspective. A morally deep world point of view would argue that both interests need to be considered carefully, including the interests of the entire park community and those of the "surplus" wolf.

A criticism of a morally deep world perspective is that it prevents any action that will affect a community. On the contrary, though a morally deep perspective does assert actions that violate vital interests of the community or erosion of it self-identity should be avoided, it requires active participation in the protection of the essential functions and the maintenance of the viability of life processes. Rather than calling for inaction, a morally deep world perspective suggests contemplation followed by a direct and specific response.

4.3   A NEW ENGINEERING ETHIC

Given a shift to a morally deep world paradigm, a new engineering code of conduct is outlined. The majority of existing codes are structured in similar if not identical ways with fundamental principles supported by fundamental canons. That same structure will be incorporated into the present work. For a morally deep world, the first fundamental canon and rule of practice is specified as:

> *Engineers, in the fulfillment of their professional duties, shall hold paramount the safety, health and welfare of the identified integral community.*

The fundamental difference between an ethical code based on a morally deep world versus the present codes is the replacement of the "public" by the "identified integral community." The important difference this substitution makes can be seen in the following set of cases.

4.4   CASE STUDY 2: A PLOW FOR MEXICAN PEASANT FARMERS

The case [54] is described as follows: "There is a pressing need for a device to assist third-world peasant farmers in cultivating their small plots of land. This need has never been satisfactorily met by any of the plows currently available. This case involves the design of a plow which can fulfill this need."

The case study went on to identify important questions concerning the significance of the plow:

- For whom should the plow be designed?

- Will humans or animals be used to pull the plow?

- Is the design of the plow sensitive to the gender of the operator?

- Will the operator of the plow walk or ride?

- Should the plow be designed at all?

After careful ethical analysis using present codes, several conclusions are reached. First, it is stated that a designer cannot answer all of the questions posed.

In order to do so, the engineer would not only have to do an enormous amount of research, but also have to know the particular social group for which the plow is being designed. Second, quite surprisingly, the stated purpose of the discussion of the case under the present codes did not intend to result in an engineer becoming "obsessed with the cultural and ethical aspects of her (sic) work that she loses sight of more narrowly engineering considerations." Third, the intent of the case study was to raise the issue of "problems of conscience" as they arise in engineering work.

In order to implement a morally deep world code, we will first have to identify members of the integral community. A listing may include:

- Peasant farmers

- Landowners

- Animals used in working the land

- Peasant society and culture

- Health of the land

- Indigenous society and culture

- Local society and culture

With such identified members of the community, the first two conclusions are very different. Whether or not a great deal of research on the various social groups is no longer an option for the engineer but required. Secondly, leaving aside "obsession", the engineer is required to be aware of the cultural and ethical aspects involve or technical aspects in the proposed design as well as the more narrowly defined engineering

4.5 CASE STUDY 3: A TICKET TEARING DEVICE FOR A DISABLED PERSON

The case [55] is described as follows: David is a young man who suffers from a variety of physical and metal disabilities. David was employed at a movie theater in his local community near Philadelphia. His primary responsibility was to welcome patrons as they went into the theater hall, taking their admission tickets, tearing them in half and placing the torn tickets into a receiving basket. As David had very limited strength in his hands, the lines of people seeking admittance would soon back up. It was determined by both his employer and social worker that some variety of a device that would help David's pace would be a great aid. A team of senior engineering capstone design students selected this project and dedicated two semesters to the design, fabrication, testing evaluation and delivery of the final device.

During the two terms, David made several visits to the campus and he and the students became quite close. Delivery day became a highly publicized event with local officials, university officials, family and friends all in attendance along with local and national press. David thoroughly enjoyed the festivities and was immensely pleased by his device. At that time, the project seemed an incredibly successful effort for everyone. Subsequent to the celebration, David continued his work for a while as an attendant at the theater but soon things began to change. He became much more withdrawn than he had ever been and soon quit his job. The seeming depression became worse and worse notwithstanding the heroic efforts of his social worker. David now is completely withdrawn and in fact institutionalized.

For this case, the following questions concerning the significance of the ticket-tearing device can be asked:

- For whom should the ticket-tearer be designed? David? Or a generic client?

- Who will provide technical support for David after the project is finished?

- Who will provide emotional support for David after the project is finished?

- Is the design of the ticket-tearer sensitive to the particular situation of David?

- Is the entire project sensitive to the particular situation of David?

- Should the ticket-tearer be designed at all?

The identified members of the community may include:

- David

- David's family

- Technical support workers

- Counseling support

- Theater owner

- Theater patrons

- The physically and emotionally disabled

An objective judging of this case would clearly point to the fact that notwithstanding all the noblest of intentions, David is now worse off than ever before. An engineering team though they followed all appropriate engineering dictums of safety, durability, etc., delivered a device that ultimately may have contributed to the suffering of a young man. What if instead of the engineering codes of conduct and ethics in place today, a morally deep world approach is taken? What would the consequences of such an approach be in this particularly poignant case study?

I would suggest that such an approach would force a much broader consideration of all the factors at play in David's case. There would be consideration given to the impact of not only the device but also the associated attention that the project garnered, an integration of many more professional perspectives, a consideration of not only short term benefits but also those of a much longer time scale. A consideration of David's family and friends and their support for him would be factored into the design.

4.6   CONCLUDING REMARKS

Vesilind [56] in his recent book, *The Right Thing to Do*, rightly points out that the code of ethics is a "fine, first but rough tool for making decisions in engineering." Vesilund also states that the present codes have virtually nothing to say about the environment nor do they spell out what, if any, responsibilities engineers have to non-human animals, plants or places. The present review of existing codes points to the validity of this assertion. A review of presently available case studies for engineering education also struggle with issues associated with social, and cultural impacts. The present work offers one approach to insuring each of these considerations is included in the engineering approach. The approach is based on the notion of a morally deep world first developed within the context of environmental ethics. The key to implementation of this approach is in the identification of key elements in what was referred to as the integral community, that is, both the different elements of a community in an analytical sense but also the community, as a whole, synthetically.

The present codes of ethics used in engineering date back to a period when the universe and all its parts were thought to be nothing more than a great and grand machine. Our understanding of the universe and the natural world here on Earth has dramatically changed since that era. Now, we speak of emergent properties in self-organized systems. I would suggest that it is time for us as engineering educators and engineers to consider adopting a new code of ethics based on our new understanding.

CHAPTER 5

Engineering Design in a Morally Deep World

Engineering design is a process that creates and transforms ideas and concepts into a product definition that satisfies customer requirements. The role of the design engineer is the creation, synthesis, iteration, and presentation of design solutions. The design engineer coordinates with engineering specialists and integrates their inputs to produce the form, fit and function documentation to completely define the product.

5.1 OVERVIEW OF TRADITIONAL ENGINEERING DESIGN

In the traditional modern world view, engineering design and development of new products and services are considered essential for the welfare of most industries in the world [57]. During the past decade, there have been many changes to product and service development but a general outline of the approach is presented in Table 5.1 [58].

Engineers are expected to design any system, subsystem, component or process in accord with the following important design factors: (1) performance; (2) appearance; (3) manufacturing; (4) assembly; (5) maintenance; (6) environment; and (7) safety.

Comparing a traditional engineering design methodology is insightful in light of the challenges of peace and security, poverty and the underdeveloped world

TABLE 5.1: Traditional Engineering Design Algorithm

STEP	DESCRIPTION OF ACTIVITY
1	Identify the customer needs.
2	Establish the product specification.
3	Define alternative concepts for a design that meets specifications.
4	Select the most suitable concept.
5	Design the subsystems and integrate them.
6	Build and test a prototype and then improve it with modifications.
7	Design and build the production facility.
8	Produce and distribute the product.
9	Track the product after release developing an awareness of its strengths and weaknesses.

and environmental sustainability. While there is a mention of designing for the environment, the methodology explains that "customers and society at large are concerned with preserving the environment" and they also want to minimize the impact of new products and processes upon the environment. There is no mention of issues related to security nor is there any mention of the problem of poverty and the underdeveloped world suggesting that engineering and engineers need not be concerned.

5.2   ECO-EFFECTIVE DESIGN

McDonough[1] and Braungart [59] challenge the notion of minimizing the impact on the environment by asking the provocative question: "When is being 'less bad' no

---

[1]William McDonough is a world-renowned architect and designer and winner of three U.S. presidential awards: the Presidential Award for Sustainable Development (1996), the National Design Award (2004); and the Presidential Green Chemistry Challenge Award (2003). *Time* magazine recognized him as a "Hero for the Planet" in 1999.

good?" They offer a design methodology termed "cradle-to-cradle" based on eco-effectiveness, rather than eco-efficiency (i.e. being "less bad."). The characteristics of a system or product designed according to the different design philosophies are compared in Fig. 5.1.

To illustrate a process or industry which utilizes an eco-effective model of design, McDonough, and Braumgart [60] describe a community of ants. As part of the ant community's daily activities, they:

- "Safely and effectively handle their own material wastes and those of other species.

- Grow and harvest their own food while nurturing the ecosystem of which they are a part.

- Construct houses, farms, dumps, and food-storage facilities from materials that can be truly recycled.

- Create disinfectants and medicines that are healthy, safe and biodegradable.

- Maintain soil and health for the entire planet."

The cradle to cradle design methodology makes important contributions in promoting environmental awareness and sustainability. This approach is particularly focused on the important issue of respect for diversity and in stark opposition to what the authors' refer to as the "attack of the one size fits all" design response and its attendant de-evolution or simplification on a mass scale. Issues related to the challenge of peace and security and the challenge of poverty and under-development are not addressed directly though there is a reference to producing abundance of natural materials and energy sources. The questions remains—who would have access to this abundance? Would it be distributed throughout the world as wealth is distributed now? How ultimately with that distribution impact questions of peace and security? Such questions seem of pre-eminent importance and need to be more fully addressed. Engineering within the context of a morally deep philosophy offers

Comparison of design methodologies

Eco-efficiency vs. Eco-effectiveness

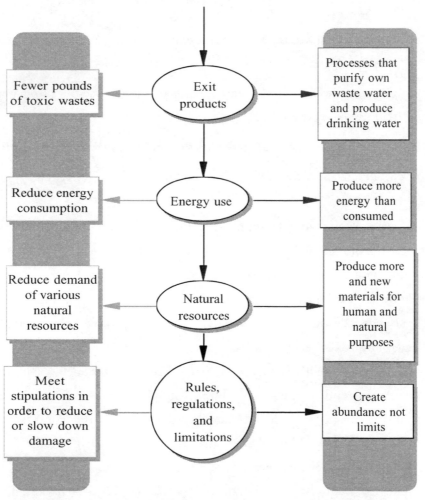

FIGURE 5.1: Comparison of eco-efficiency vs. eco-effectiveness in design

hope and a design methodology using the morally deep paradigm will be presented in the following section.

5.3   A DESIGN ALGORITHM FOR A MORALLY DEEP WORLD

A new design algorithm, which can be partitioned in the following four steps, is offered (Fig. 5.2):

- Via Positiva. The problem is identified, fully accepted and broken down into its various components using the vast array of creative and critical thinking techniques which engineers possess. What is to be solved? For whom is it to be solved?

- Via Negativa. Reflection on the possible implications and consequences for any proposed solution are explored. What are the ethical considerations involved? The societal implications? The global consequences? The effects on the natural environment?

- Via Creativa. The third step refers to the act of creation. The solution is chosen from a host of possibilities, implemented and then evaluated as to its effectiveness in meeting the desired goals and fulfilling the specified criteria.

- Via Transfomativa. The fourth and final step asks the following questions of the engineer: Has the suffering in the world been reduced? Have the social injustices that pervade our global village been even slightly ameliorated? Has the notion of a community of interests been expanded? Is the world a kinder, gentler place borrowing from the Greek poet Aeschylus? [61]

5.4   CASE STUDY 1: GRAPE WORKERS IN NORTHERN CALIFORNIA

Consider the following scenario. Harvesting machines (Fig. 5.3) are replacing migrant workers as grape pickers in northern California [62]. Today, such machines

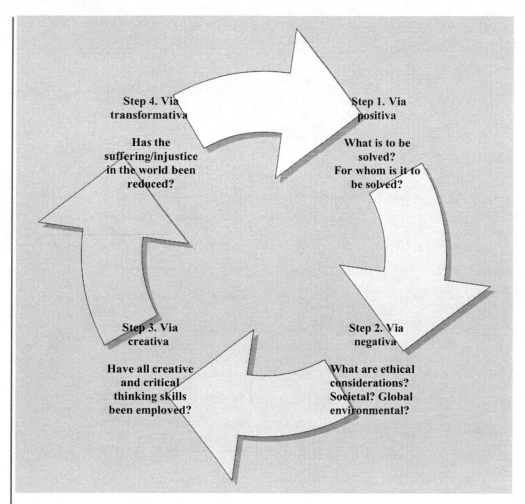

Step 4. Via transformativa

Has the suffering/injustice in the world been reduced?

Step 1. Via positiva

What is to be solved? For whom is it to be solved?

Step 3. Via creativa

Have all creative and critical thinking skills been employed?

Step 2. Via negativa

What are ethical considerations? Societal? Global environmental?

FIGURE 5.2: Representation of an engineering design algorithm based upon a morally deep world

cost approximately $126 000. It takes five workers about eight hours to pick 10 tons of grapes. To harvest the 100 tons would require 90 workers if they wanted to be done before sun-up. The costs are relatively straightforward to calculate. It costs $120 a ton to harvest by hand, and $30 a ton by machine. A mid-size wine producer expects to harvest about 4000 tons of grapes total. For 4000 tons, the cost to pick the grapes by hand is approximately $480 000 while the costs using a harvester are

FIGURE 5.3: An example of a mechanical fruit picker [63]

$120 000 representing an increase in margin above cost equal to $340 000. Even with the cost of the harvester subtracted from this margin, the net increase in profits is equal to $204 000.

Traditional engineering ethics discussions might stop here. But what if we go a bit farther and identify the workers as being part of the community? What else would we be forced to consider? If it takes 90 workers to harvest 100 tons then it would take the equivalent of 3600 workers to harvest 4000 tons. Certainly there are many more people involved than simply the workers themselves with the total involved growing far beyond the estimate of 3600. What will become of them? Should we as the engineers who designed, built, tested, evaluated and delivered the harvester care? If we use the code of ethics described by countless engineering societies today, the answer would be no. There is no consideration given to the workers whose livelihoods have been eliminated. Codes of engineering ethics based on the notion of a morally deep world would suggest a very different result as it forces us as engineers to consider the entire living system, in this case, including the men, women and children who toil as grape pickers and whose quality of life is intimately linked to the harvest.

For this case, the following questions concerning the significance of the grape-harvesting device may be asked:

- For whom should the device be designed? The owners? The land? The workers? Someone else?
- What will become of the displaced workers?
- What will become of the displaced workers families?
- Are there long-term effects on the land?
- Are there societal implications for the workers' communities?
- Are there societal implications for the greater community outside the workers and their families?

The identified members of the community may include:

- Owners
- Workers
- Workers' families
- Land and water ecosystem
- Farm worker society
- Local, regional and national societies

Suppose we use the proposed engineering design algorithm. In Via Positiva, an identification of the problem is made. Grapes have to be picked at a particular time and in a particularly rapid fashion. The harvesting of grapes by hand has in fact been done satisfactorily for thousands of year on farms of all sizes and now to a limited extent by machines. In Via Negativa, the possible implications or consequences of the change to a harvesting machine are considered. The effects upon the environment for the given case seem minimally affected by the mechanization of the process as stated here. The consequences for the workers, their families and their ways of life are however profound and disturbing. A large number of workers will be displaced and their jobs completely eliminated. In Via Creativa, engineers have in fact designed and delivered machines that will meet the criteria set forth by the grape farmers and have thus met their professional responsibilities as has been commonly understood. But a consideration of the net effects on the suffering in the world, Via Transformativa, might yield a different result. Engineers who design such devices without concern for the impact on the numbers of farm workers who are being replaced solely in order to increase the profits of the landowners may have acted unethically.

In the context of the farm workers discussed here, a logical question to ask would be the following: Is it possible to arrive at an end result which was a creative design that met everyone's needs and established justice for workers and their families and furthered the interests of the vineyard? While we cannot offer a device, we can suggest that those whom the new machine would make irrelevant might be included in discussions concerning the consequences of implementing the new design. The landowner may ultimately decide to mechanize this task but at the very least the criteria for making ethical choices would also include a careful consideration of the impact of the engineering design on the lives of the workers. It would seem such an inclusion is as important as any other.

5.5 CASE STUDY 2: TRANSPORTING TOURISTS IN CAPE CHURCHILL

Consider the following case of a dune buggy designer whose clients want a vehicle suitable for taking tourists out along the shore of Hudson Bay to view polar bears. Suppose a company wishes to increase tourism in remote Churchill by promoting the opportunity to view polar bears in an intimate setting. To do this a new version of the ubiquitous dune buggy will be required, one that can withstand the harsh

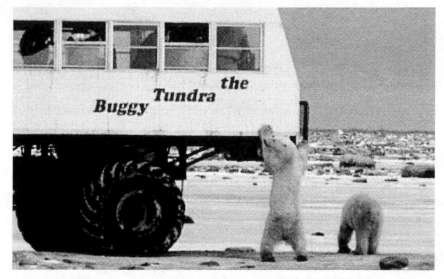

FIGURE 5.4: Tourists and polar bears interact near Cape Churchill [64]

climate, rough terrain and stand up to an occasional overly curious polar bear. Suppose after much deliberating, the dune buggy's conceptual design is complete. The vehicle shall be fully enclosed and heated and be capable of transporting approximately 25 people at a time. Presumably if tourists can view polar bears in this relatively comfortable and safe setting, more tourists will be drawn to Cape Churchill (Fig. 5.4).

For this case, the following questions concerning the significance of the dune buggy device may be asked:

- For whom should the device be designed? The owners? The tourists? The polar bears? Someone else?

- With more tourists, what will be the short-term effects on the local community? The local indigenous community? The polar bears? The tundra?

- With more tourists, what will be the long-term effects on the local community? The local indigenous community? The polar bears? The tundra?

- What are the implications of promoting eco-tourism?

The identified members of the community may include:

- Owners

- Towns people

- Indigenous peoples

- Polar bears

- Ecosystem in Cape Churchill

- Local, regional and national communities

Whether or not the Arctic-compatible version of a dune buggy is designed and delivered (in this case, it was) would depend upon much more than the needs of a few owners in a morally deep world-view. In the second step of the design process, Via Negativa, attention would be paid to careful considerations of the societal and environmental implications of the device. In the third step of the design process, Via Transformativa, the actual question of whether or not ecotourism is a positive or negative factor for the health of a fragile ecosystem as is the one at Cape Churchill would be raised. In promoting more contact with the polar bears of Cape Churchill are we condemning them and their species to a more rapid extinction? Or are we helping them by bringing a greater awareness to a wider audience?

5.6   CONCLUDING REMARKS

An engineering design algorithm based on a morally deep world-view has been developed. I have considered several cases in a careful examination of the implications of using the new algorithm. I have looked at one case, which focuses primarily upon societal implications while the other focuses upon the possible impact on one particular ecosystem, the Arctic tundra, and one particular species, polar bears.

As with all other attempts at integrating an ethical foundation into the engineering profession, it is not the particular ethical theory, which provides a correct answer, as ethical dilemmas do not have right and wrong answers. If such answers existed, we would simply not use the term dilemma.

A morally deep world-view integrated into our approach to design does afford us something very important as engineers. It allows us to consider a much broader range of "clients", many more than our simply paying our salaries. It also provides us with a mechanism whereby we can refuse ethically to work on a particular project even though the device itself may meet basic safety requirements. We also may refuse to work on projects that held paramount public safety but excluded considerations of the environment or the impact our device might have upon a community.

CHAPTER 6

Implications for Engineering Education

An engineering profession that defines ethics using a morally deep paradigm would, in turn, require changes to the manner in which students are educated. In the present chapter, an educational paradigm based on the notion of an integral community is described. Engineering education is highly regulated by accrediting agencies, particularly in the United Sates. In many ways, it is the accreditation process that is most effective in promoting change in engineering education. As a result, changes are suggested for the criteria used by the Accreditation Board in Engineering and Technology (ABET) in determining whether or not any program is fully sanctioned.

6.1 A NEW PARADIGM FOR ENGINEERING EDUCATION

In June 2000, an international conference entitled "Connecting Ethics, Ecological Integrity and Health in the Millennium" included one particular presentation, which focused upon what was termed by the authors a peace paradigm for education [65]. The model, the Integral Model of Education for Peace, Democracy and Sustainable Development, adopted from the Earth Charter [66] described three components needed in the shift (Fig. 6.1). The three components describe the three fundamental interrelationships, which form the contexts of our lives: living with ourselves, living with others, and living with nature. Each of these three

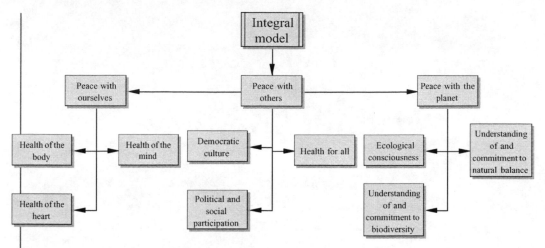

FIGURE 6.1: Integral model of education for peace, democracy and sustainable development

relationships can be characterized as being marked by violence or by peace. In turn, each of the three relationships can be further subdivided into different dimensions or levels of expression.

6.1.1   Living in Peace with Ourselves

The interrelationship, living at peace with oneself, is subdivided into levels of expression corresponding to living at peace in the body, in the heart, and in the mind. Having peace in our body entails the development of an awareness or consciousness of our physical needs for health and a wise optimization of any/all satisfiers of those needs. This may be seen as an indictment of the "all-nighter" work ethic that is far too common-place among engineering majors. (On a personal note, appealing to students' health and well-being has not been particularly effective for us in teaching time management skills. Perhaps it may be more effective to link the importance of such skills to the promotion of peace.) In addition, being at peace with ourselves certainly points to issues of proper nutrition and substance abuse among others. Peace in our heart requires the meeting of needs that generate a sense of basic security and trust. Here it is necessary to cultivate qualities such as love, compassion and tolerance. Peace in our mind refers to the possibility of self-realization based on an ethical consciousness of universal responsibility, that is, an appreciation of one's place in natural and human history, and understanding the interdependence of all beings in the universe, as well as the present day global challenges.

6.1.2 Living in Peace with Others

Living in peace with others has the main themes of a truly democratic culture, political and social participation, and health for all. Democratic culture encompasses the key elements of meaningful participation in societal affairs, a sense of responsibility and a sense of solidarity. Elements involved in political and social participation include a fully engaged and participating citizenship, an identification of the common good and an understanding of the principles of peaceful conflict resolution. What is not called for is an insistence upon a particular implementation of democracy; rather there is a recognition and respect that different societies and cultures may reach decisions in a fully democratic way. As an example, we offer the practice of reaching consensus and decisions practiced by our neighbors, the Haudenosaunee, who live on the Onondaga Nation territory in upstate New York. The Onondaga Nation government consists of the traditional Council of Chiefs and Clan Mothers. Additionally, the Onondaga Chiefs sit on the Haudenosaunee Grand Council. Chiefs from each of the Six Nations meet regularly at Onondaga. This governmental organization served as the actual model used by the founders of the American republic [67]. Lastly, the third level or component, health for all, requires an understanding and practice of generosity, an understanding of being itself as a guide for having and doing and a sense of economic security for all society with an absence of a fear of scarcity.

The translation of living in peace with others to the goals of democracy, political participation and health may be questioned in the following way: Are these truly universal values or simply the highest values embraced by Western civilization? How can we in the West be sure that we are not arbitrarily imposing these goals on non-Western cultures in order to remake their societies into versions of our own? We would suggest that though these are difficult questions to answer, perhaps the most important contribution we can make as educators is provide a forum within which students may wrestle with these issues.

6.1.3 Living in Peace with the Planet

Included in the concept of living at peace with the planet are an ecological consciousness, an understanding and commitment to biodiversity as well as an understanding and commitment to the maintenance of a natural balance. The use of the word "peace" to describe environmental care/stewardship is intentionally provocative. In my view the notion of care and stewardship presupposes a hierarchical relationship between humankind and the planet whereas peace summons forth a metaphor of sitting at a bargaining or "peace" table as equals. In turn, each of these three elements is further subdivided into three components. Ecological consciousness entails identity with the cosmos, an understanding and respect for evolutionary forces and ultimately a respect for life. Here the respect for evolutionary forces does not refer to the narrow view of "survival of the fittest." Rather it calls on us to recognize that nature is ever changing, ever recasting itself, and not to seek to halt or reverse all such changes. This does not preclude efforts to protect endangered species per se but it does call for a careful consideration prior to intervention. Perhaps, after reflection, society may wish to preserve gray wolves or Bengal tigers through government action while not permitting continued dredging of the Atchafalaya River Basin near New Orleans, Louisiana. Biodiversity consists of an appreciation for the place in the web of life of the various plants and animals, a commitment to the protection of species, particularly endangered species and a commitment to conservation in concert with the dynamic nature of ecosystems. Natural balance encompasses an appreciation of the integrity of natural systems, an emphasis on sustainable resource use as well as on the importance of ecological security.

6.2 ACCREDITATION CODES AND MODIFICATIONS

In the United States, engineering programs are tightly regulated by the Accreditation Board for Engineering and Technology (ABET) [68]. ABET, Inc., is the recognized U.S. accreditor of college and university programs in applied science, computing, engineering, and technology. ABET was established in 1932 and is now a federation of 28 professional and technical societies representing the fields of applied science, computing, engineering, and technology. According to ABET, their accreditation process ensures the quality of the postsecondary education students receive. Educational institutions or programs volunteer to periodically undergo this review in order to determine if certain criteria are being met. The accreditation is not a ranking system; rather it is simply assurance that a program or institution meets established quality standards.

Accreditation is a voluntary process on the part of an institution. The steps in the accreditation process are: the engineering program performs a self-study, a team of outside ABET evaluators then visit campus, conduct interviews and compare the program with established criteria, and a final report and finding are returned back to the institution for their review.

ABET sets forth a set of criteria to be met in securing accreditation [69]. The set of requirements cover the quality of students, program educational objectives, programs outcomes and assessment, professional component, the quality of the faculty, classroom and laboratory facilities, institutional support and financial resources, and program criteria. Considering the existing ABET criteria, Criterion 3 focuses upon program outcomes as shown below. The modified Criterion 3 incorporating the integral model with the changes typed in bold, italics may be written as the following:

Engineering programs must demonstrate that their graduates have:

a) An ability to apply knowledge of mathematics, science, and engineering;

b) An ability to design and conduct experiments, as well as to analyze and interpret data;

c) An ability to design a system, component, or process to meet desired needs;

d) An ability to function on multidisciplinary teams;

e) An ability to identify, formulate and solve engineering problems;

f)   An understanding of professional and ethical responsibility;

g)   An ability to communicate effectively;

h)   The broad education necessary to understand the impact of engineering solutions in a global and societal context;

i)   A recognition of the need for, and an ability to engage in life-long learning;

j)   A knowledge of contemporary issues;

k)   An ability to use techniques, skills, and modern engineering tools necessary for engineering practice.

The modified Criterion 3 incorporating the integral model would include the following additional program objective (l):

A fully integrative approach to engineering problems incorporating both reason and compassion in the development of solutions.

6.3   CONCLUDING REMARKS

A new model for engineering education incorporating a morally deep world view and the notion of an integral community is presented. The model is based upon making and nurturing peace with ourselves, with others and with the planet. In addition, as engineering education is highly regulated through the action of accreditation agencies such as ABET, a modification is proposed to the existing ABET Criterion III which supports the integral model by requiring an incorporation of both reason and compassion in the development of proposed technical solutions.

CHAPTER 7

Final Thoughts

I began this discussion with an identification of the three most pressing challenges we face as a profession in this new millennium: the challenge of security, the challenge of poverty and underdevelopment and the challenge of environmental sustainability. The profession of engineering has many important connections to each of the separate challenges. We take great pride in the strong ethical foundation of our profession and have gone to great lengths to try to codify the important ethical dimension of our profession.

Yet, it is here where I think we have fallen short. The ethical codes put forward by countless engineering societies and engineering education agencies are by and large locked into a world-view that was first developed in the Age of Enlightenment. The science of that period in our history was dominated by a view of Nature and the Universe as a grand machine. They, the scientists and philosophers of the time, spoke of Nature as a mechanical clock. As is always the case, science has changed, and now the current paradigm is one of Nature as a self-organizing system with seemingly chaotic systems giving rise to structure through emergent properties. I have offered a different ethical code based upon the science of today as first applied to environmental ethics by Johnson. In his morally deep world-view, Johnson described what he referred to an integral community. It is, I believe, the integration of the notion of an integral community into engineering ethics that holds great promise for the future of our profession.

Consider its implications. When challenged to develop a new device or system, we can now consider a much broader range of "clients" and responsibilities.

To use a visual metaphor, we can now imagine we are dropping a small pebble into a still pond and watch the waves propagate ever outward from the point of impact. We can imagine these waves traveling farther and farther, as far in fact, as our sense of responsibility will take us. I would further offer that once we begin to enlarge the area encompassed by that wave, we will be drawn to enlarge it even more.

As engineers, we have helped develop weapons systems that would only a few years ago seemed more fitting a Jules Verne novel. So too have we helped create incredible wealth for a small portion of the world's community. The impacts of many of our technological advances have been devastating for the world's ecosystems, the species of plants and animals and the indigenous peoples who make up the largest percentage of the world's population. However well intentioned our codes of ethics in the past are, it is my view that they have not met the challenges we have identified for us now in the 21st century nor will they serve us well as we move inexorably into the future.

References

[1] "State of the World 2005," *A Worldwatch Institute Report On Progress Toward a Sustainable Society*, New York: W.W. Norton & Company, 2005, pp. xvii–xxi.

[2] Ernest Greenword. "Attributes of a Profession," *Social Work*, pp. 45–55, July 1957.

[3] "Psychiatrist Summoned to Moscow to Help Ease Suffering in Breslan, Russia," *Medical News Today*, 10 March 2006, http://www.medicalnewstoday.com/medicalnews.php?newsid=14033.

[4] "BBC News In Depth," http://news.bbc.co.uk/1/hi/in_depth/europe/2004/madrid_train_ attacks/default.stm.

[5] "Militarism and Armed Conflicts," *World Centric*, http://www.worldcentric.org/stateworld/military.htm.

[6] H. McDonald. "Origins of the Word 'Engineer,'" *ASCE Transactions*, vol. 77, p. 1737. Reproduced in ASCE Committee on History and heritage of Amcrican Civil Enginccring, 1970, Historical Publication No. 1.

[7] Aarne Vesilind. *Peace Engineering*, Woodsville, NH: Lakeshore Press, 2005, pp. 1–2.

[8] "Social and Economic Justice," *World Centric*, http://www.worldcentric.org/stateworld/military.htm.

[9] "The Preamble," *Program of Action of the United Nations International Conference on Population & Development*, http://www.iisd.ca/Cairo/program/p01000.html.

[10] "Chapter II-Principles," *Program of Action of the United Nations International Conference on Population & Development*, http://www.iisd.ca/Cairo/program/p02016.html.

[11]   Daniel A. Vallero. "Just Engineering: Peace, Justice, and Sustainability," *Peace Engineering*, Woodsville, NH: Lakeshore Press, 2005, pp. 41–56.

[12]   Krista E.M. Galley, ed. *Global Climate Change and Wildlife in North America*, Technical Review 04-2, Bethesda, MD: The Wildlife Society, Dec. 2004.

[13]   AP Wire. *Ocean Dead Zones on the Increase*, CBS News, 10 March, 2006, http://www .cbsnews.com/stories/2004/03/29/tech/main609151.shtml.

[14]   Lori Brown. "State of the World: A Year in Review," *State of the World 2005*, Norton, New York: Worldwatch Institute, 2005, pp. xxiii–xxvii.

[15]   Susan Joy Hassol. *Impacts of a Warming Arctic*, New York: Cambridge University Press, Dec. 2004.

[16]   Brian Handwerk. "Global Warming Could Cause Mass Extinctions by 2050, Study Says," *National Geographic News*, http://news.nationalgeographic.com/news/2006/04/0412_060412_global_warming.html.

[17]   Jon Copely. "The Great Ice Mystery," *Nature*, vol. 408, pp. 634–636, 07 Dec. 2000.

[18]   "African Farmlands Plagued by Severe Degradation, Study Finds," *The Foundation Directory Online*, The Foundation Center, http://foundationcenter.org/pnd/news/story.jhtml;jsessionid=S3ICALZI0G2OPTQRSI4CGW15 AAAACI2F?id=138800010.

[19]   "Webster's On-line Dictionary," www.webster-dictionary.org/definition/engineering.

[20]   "Ethics," http://en.wikipedia.org/wiki/Ethics.

[21]   Jeffrey Wattles. *The Golden Rule*, Oxford: Oxford University Press, Dec. 1996.

[22]   Ethics, *The Internet Encyclopedia of Philosophy*, http://www.utm.edu/research/iep/e/ethics.htm.

[23]   Kant's Moral Philosophy. *Stanford Encyclopedia of Philosophy*, http://plato.stanford.edu/ entries/kant-moral/.

[24]   W.D. Ross. *The Right and Good*, http://www.ditext.com/ross/right.html.

[25]   "Consequentialism," http://www.utm.edu/research/iep/c/conseque.htm.

[26] Jeremy Bentham. *The Classical Utilitarians: Bentham and Mill*, Indianapolis, IN: Hackett Publishing, Mar. 2003.

[27] G.E. Moore. *Principia Ethica*, http://fair-use.org/g-e-moore/principia-ethica/.

[28] *NSPE Code of Ethics*, National Society for Professional Engineers, http://nspe.org/ethics/ eh-1code.asp.

[29] *ASME Code of Ethics of Engineers*, American Society of Mechanical Engineers, http:// asme.org/asmeorg/Governance/5431.doc.

[30] *ASCE Code of Ethics*, American Society of Civil Engineers, http://www.asce.org/inside/codeofethics.cfm.

[31] *ASCE Code of Ethics*, American Society of Civil Engineers, http://www.asce.org/inside/ codeofethics.cfm#note3.

[32] *IEEE Code of Ethics*, The Institute of Electrical and Electronics Engineers, http://www .iee.org/portal/site/mainsite.

[33] *IIE Code of Ethics*, The Institute of Industrial Engineers, http://www.iienet.org/public/ articles/details.cfm?id=79.

[34] "2006–2007 Criteria for Accrediting Engineering Programs, Accreditation Board for Engineering and Technology," http://www.abet.org/Linked% 20Documents-UPDATE/Criteria%20and%20PP/E001%2006-07% 20EAC%20Criteria%2012-19-05.pdf.

[35] *AIChe Code of Ethics*, American Institute of Chemical Engineers, http:// www.aiche.org/ About/Code.aspx.

[36] David Hume. *A Treatise of Human Nature* (*Oxford Philosophical Texts*), New York: Oxford University Press, 2000.

[37] Gail Dawn Baura. *Engineering Ethics, 1st Edition: An Industrial Perspective*, New York: Academic Press, 2006.

[38] Christopher Gabbard. *Gender and Politics in Literature, 1688–1750*, http://www.stanford.edu/class/engl174b/main_page.html.

[39] Lewis Mumford. *Technics and Civilization*, New York: Harvest/HBJ Books, 1963.

[40]   Umberto Eco. *The Open Work*, Cambridge, MA: Harvard University Press, 2005.

[41]   Toby Huff. *An Age of Science and Revolution, 1600–1800*, New York: Oxford University Press, 2005.

[42]   James E. Lovelock. *Homage to Gaia: The Life of an Independent Scientist*, New York: Oxford University Press, 2001.

[43]   James E. Lovelock. *Gaia: A New Look at the Earth*, New York: Oxford University Press, 2000.

[44]   Erich Jantzt. *Self Organizing Universe: Scientific and Human Implications*, New York: Pergamon Press, 1980.

[45]   F. Eugene Yates, ed. *Self-Organizing Systems: The Emergence of Order*, Springer Verlag, 1988.

[46]   Ethan Decker. *Self Organizing Systems: A Tutorial in Complexity*.

[47]   *Global Radiation Patterns: December–January 1991*, National Oceanic and Atmospheric Administration, http://www.noaa.gov/and http://www.holon.se/folke/images/Globrad1.jpg.

[48]   Lawrence Johnson. *A Morally Deep World: An Essay on Moral Significance and Environmental Ethics*, Cambridge: Cambridge University Press, 1993, pp 12–50.

[49]   Aldo Leopold. *A Sand County Almanac*, New York: Ballantine Books 1986.

[50]   Lawrence Johnson. *A Morally Deep World: An Essay on Moral Significance and Environmental Ethics*, Cambridge: Cambridge University Press, 1993.

[51]   Catharine Feher-Elston. *Wolfsong*, New York: Tarcher Press, 2005.

[52]   Doug Smith and Gary Ferguson. *Decade of the Wolf: Retuning the Wild to Yellowstone*, Guilford, CT: The Lyons Press, 2005.

[53]   Gary Wockner, Gregory McNamee, and Sue-ellen Campbell. *Comeback Wolves*, Boulder, Colorado: Johnson Books, 2005.

[54]   "Engineering Ethics Case Studies," http://ethics.tamu.edu/ethics/plow/plow.htm.

[55] George D. Catalano. "Compassion Practicum: A Design Experience at the United States Military Academy," *ASEE Journal of Engineering Education*, 2001.

[56] P. Aarne Vesilund. *The Right Thing to Do: An Ethics Guide for Engineering Students*, New Hampshire: Lakeshore Press, 2004.

[57] Clive Dym and Patrick Little. *Engineering Design: A Project Based Approach*, New York: Wiley, Nov. 2003.

[58] James Daly and David Rocheleau. *Introduction to Engineering Design*, Knoxville, TN: College House Enterprises, LLC, 1998.

[59] Michael McDonough and Michael Braungart. *Cradle to Cradle: Remaking the Way We Make Things*, New York: North Point Press, 2002, pp. 45–67.

[60] Michael McDonough and Michael Braungart. *Cradle to Cradle: Remaking the Way We Make Things*, New York: North Point Press, 2002, pp. 121–122.

[61] Aeschylus, Herbert Weir Smith. *Aeschylus*, Cambridge, Massachusetts: Harvard University Press, 1970.

[62] Tim Tesconi. "California Grape Harvest 2000," *The Press Democrat*, Oct. 26, 2000.

[63] "Harvesting an Olive Orchard," http://www.oliveoilsource.com/harvesting .htm.

[64] "The Tundra Buggy Adventure," http://www.tundrabuggy.com/polar-bear-tours/cape-churchill/.

[65] Leonardo Boff. *Cry of the Earth, Cry of the Poor*, New York: Orbis Press, 1997.

[66] Anita Wenden. *Educating for a Culture of Social and Ecological Peace*, New York: State University of New York Press, Nov. 2004.

[67] Alexander Ewen and Chief Oren Lyons. *Voice of Indigenous Peoples: Native American People Address the United Nations*, October 1993.

[68] *Accreditation Board of Engineering and Technology*, www.abet.org.

[69] *Accreditation Policy and Procedures Manual*, ABET Board of Directors, pp. 2–21.

Author Biography

George D. Catalano is a Professor of Mechanical Engineering at the State University of New York at Binghamton. He holds joint appointments in the Departments of Mechanical Engineering and Bioengineering. In addition, he serves as the director of the university-wide honors program at Binghamton. Dr. Catalano earned a Doctor of Philosophy and Master of Science degrees in aerospace engineering at the University of Virginia and a Bachelor of Science degree also in aerospace engineering at Louisiana State University. Prior to his present position, he has served on the faculty at the Air Force Institute of Technology, Wright State University, Louisiana State University, and the United States Military Academy at West Point. He also served as a visiting scholar at the Politechnic in Torino, Italy and at the Technical Institute in Erlangen, Germany.

Dr. Catalano's research interests include turbulent fluid flows, low and high speed aerodynamics and experimental methods in physics, modeling ecosystems as well as learning strategies and paradigms, engineering ethics, engineering design and environmental ethics. He is listed in the Philosopher's Index for his published work in animal rights and environmental ethics. Dr. Catalano has over 150 technical and educational publications and has twice been selected as a Fulbright Scholar in recognition of his work in turbulent fluid mechanics.